ON THE HABITS OF THE BUTTERFLIES OF THE AMAZON VALLEY

(1853)

BY

ALFRED RUSSEL WALLACE

British Library Cataloguing-in-Publication Data
A catalogue record for this book is available from the
British Library

Alfred Russel Wallace

Alfred Russel Wallace was born on 8th January 1823 in the village of Llanbadoc, in Monmouthshire, Wales.

At the age of five, Wallace's family moved to Hertford where he later enrolled at Hertford Grammar School. He was educated there until financial difficulties forced his family to withdraw him in 1836. He then boarded with his older brother John before becoming an apprentice to his eldest brother, William, a surveyor. He worked for William for six years until the business declined due to difficult economic conditions.

After a brief period of unemployment, he was hired as a master at the Collegiate School in Leicester to teach drawing, map-making, and surveying. During this time he met the entomologist Henry Bates who inspired Wallace to begin collecting insects. He and bates continued exchanging letters after Wallace left teaching to pursue his surveying career. They corresponded on prominent works of the time such as Charles Darwin's *The Voyage of the Beagle* (1839) and Robert Chamber's *Vestiges of the Natural History of Creation* (1844).

Wallace was inspired by the travelling naturalists of the day and decided to begin his exploration career collecting specimens in the Amazon rainforest. He explored the Rio Negra for four years, making notes on the peoples and languages he encountered as well as the geography, flora, and fauna. On his return voyage his ship, Helen, caught fire and he and the crew were stranded for ten days before being picked up by the Jordeson, a brig travelling from Cuba to London. All of his specimens aboard Helen had been lost.

After a brief stay in England he embarked on a journey to the Malay Archipelago (now Singapore, Malaysia, and Indonesia). During this eight year period he collected more than 126,000 specimens, several thousand of which represented new species to science. While travelling, Wallace refined his thoughts about evolution and in 1858 he outlined his theory of natural selection in an article he sent to Charles Darwin. This was published in the same year along with Darwin's own theory. Wallace eventually published an account of his travels *The Malay Archipelago* in 1869, and it became one of the most popular books of scientific exploration in the 19th century.

Upon his return to England, in 1862, Wallace became a staunch defender of Darwin's landmark work *On the Origin of Species* (1859). He wrote responses to those critical of the theory of natural selection, including 'Remarks on the Rev. S. Haughton's Paper on the Bee's Cell, And on the Origin of Species' (1863) and 'Creation by Law' (1867). The former of these was particularly pleasing to Darwin. Wallace also published important papers such as 'The Origin of Human Races and the Antiquity of Man Deduced from the Theory of 'Natural Selection" (1864) and books, including the much cited *Darwinism* (1889).

Wallace made a huge contribution to the natural sciences and he will continue to be remembered as one of the key figures in the development of evolutionary theory.

Wallace died on 7th November 1913 at the age of 90. He is buried in a small cemetery at Broadstone, Dorset, England.

ON THE HABITS OF THE BUTTERFLIES
OF THE AMAZON VALLEY
(1853)

As the portion of South America, watered by the river
Amazon, has recently contributed so many new and beautiful
species of butterflies to our Cabinets, it is thought that a
few remarks on the habits of such of them as fell under the
writer's notice during a residence of four years in the country
may not be unacceptable to the Entomological Society.

It may be as well, by way of preface, to give some brief
account of the district in question.

South America, as a whole, may be looked upon as
consisting of three elevated regions, connected by intervening
low lands and valleys. Along the west side of it stretches out
from one extremity to the other the mighty Andes, though
not the loftiest, yet undoubtedly the most extensive unbroken
mountain range on the earth. Various spurs and branches
are attached to this chain, but they never extend far across
the continent in an eastern direction. Towards the Atlantic
rise the mountains and table lands of Brazil and Guiana,
two large isolated masses, everywhere separated by a wide
interval from each other and from the Andes. In the space
between these the Amazon rolls its mighty flood through a

7

vast alluvial valley, which is everywhere clothed with dense forests of lofty timber trees. The whole of this valley lies in the very centre of the tropics, and enjoys a climate in which a high and uniform temperature is combined with a superabundance of moisture.

These seem to be the conditions most favourable to the development and increase of Lepidopterous insects, and we accordingly find the valley of the Amazon to be more productive of the diurnal species than perhaps any other part of the world. Where else in a single locality can 600 species of butterflies be obtained? and this can be done within a walk of the city of Pará.

We may divide the country into four sorts of hunting grounds, each of which possesses its peculiar species rarely found in the others. 1st. The open grounds, or "campos," natural or artificial, and dry and barren places. Here we find scarcely any peculiar Amazonian species, and generally only a few of the commonest South American insects, such as *Papilio Polydamas, Danais Archippus*, and several common *Callidryas*. 2nd. The margins of the rivers, particularly during the season when the waters are falling. Here are found numerous beautiful and rare *Papilios* and species of *Timetes, Cybdelis, Callianira* and *Megistanis*, with occasionally *Aganisthos, Marpesia* and *Victorina*. 3rd. The second growth woods, plantations, and shady roads and paths. In such places the *Heliconidæ* abound, and that beautiful group of

American *Papilios* with red spotted lower wings. Here are hundreds of *Erycinidæ* and gorgeous *Theclas*; and here the *Epicalias* and *Callitheas* are to be met with, while numbers of *Morphos* flap lazily along, and *Hesperidæ*, sometimes as large as *Sphinxes*, dart by with the velocity and sound of humming birds. And, 4th, is the gloomy virgin forest, in whose damp recesses are found numbers of beautiful *Satyridæ*, many delicate *Ithomias*, and some of the most beautiful *Eurygonas*, *Mesosemias* and *Theclas*. The individuals are however never so abundant as in more open places, and much perseverance is required in the collector, which, however, seldom fails at length to be rewarded by some rare and exquisitely beautiful insect.

I shall now notice in succession the most important genera and species, with such observations on their habits as I am enabled to offer.

Beginning with the true *Papilios*, the first that presents itself is the well-known American swallow-tail *P. Protesilaus*. This is abundant in all parts of the Amazon district, but is always seen on the water's edge, where it assembles by scores and even hundreds in dense masses, preferring places where some decaying animal matter or excrement has laid,--the juices from which it sucks up with great avidity. When thus occupied I have often sat down within a few feet of them, and have caught with my fingers a single specimen of the rare *P. Agesilaus* from among a hundred of the commoner species. At such times I have noticed them ejecting from the anus a watery liquid in successive jets, which they will continue to do for a very long time. I have dissected dozens of these insects in the hope of finding a female, which I never succeeded in doing, every specimen of *Papilio Protesilaus* which I have seen being a male.

In the same places, and often mixed with the last, are found *P. Dolicaon* and the more beautiful *P. Columbus*, the latter being the more abundant of the two. In similar situations we have *Polycaon*, *Lycidas*, *Belus* and *Pausanias*, of all of which the males only are found on the river's edge, the female of *Polycaon* (the only one yet determined) frequenting the skirts of the forest, and flying about orange trees.

Thoas, *Polydamas* and *Torquatus* are found in open grounds and about houses, localities frequented by many common South American species, which have an extensive range, and do not seem to be truly indigenous to the Amazon valley.

It is in the shady groves and in the depths of the virgin forest that the great mass of the species of *Papilio* are found. These all belong to the group which have white or bluish spots on the upper wings and red on the under. They fly weaker than the other species, and seldom appear in full sunshine. *Sesostris*, *Hierocles* and *Cutora* are among the most beautiful of these; *Echelus*, *Gargasus*, *Aglaope*, *Marcius* and *Parsodes* are abundant, and very characteristic of the Amazon district. In several of these there are remarkable differences in the colours of the sexes, which do not exist in others; while of a great proportion we are still ignorant of one or the other sex. This is partly owing to the males and females seldom frequenting the same situations; thus, while *Vertumnus* and *Sesostris* (males) are found only in the damp and shady forest, their females frequent gardens, and appear to deposit their eggs upon orange trees, round which I have taken them hovering. The larvæ of butterflies, as with us, are remarkably scarce, which is the more to be wondered at, as the perfect insects are in such profusion; and thus the most satisfactory means of ascertaining the sexes of these insects can be pursued with but little success. The only *Papilios* I was enabled to breed were *Polydamas* and *Anchisiades*, species which exhibit no marked difference between the sexes. *Triopas*, which belongs to the same group, though the red and white spots are both replaced by yellow, is the smallest and most delicate of the true *Papilios*; and its habits agree with its appearance, for it frequents the shadiest part of the forest, and flies weakly and always near the ground. It is found at Pará, and all over the Amazon district; but the closely allied *Chabrias* is a native of the interior only, both on the Upper Amazon and near the tributaries of the Orinoco. In ascending the Amazon several of these interesting modifications of form occur. The *Vertumnus* of Pará is replaced at Santarem, only 500 miles up the river, by the closely allied *Cutora* of G. R.

Gray, &c. On the Upper Amazon we have *Ergeteles* instead of the *Parsodes* of Pará. The following list shows thirty-eight species of *Papilio* in the small portion of the district yet examined. I have arranged them according to the stations they frequent, and have distinguished those which occur only in the interior.

Species of *Papilio* inhabiting the Amazon valley:--

a. Species habitually frequenting the banks of the rivers at the water's edge :—

Protesilaus ♂ ..Very abundant.
Agesilaus ♂ ....Rare, with the last.
Polycaon ♂ ....The ♀ never accompanies it in this situation.
Dolicaon ♂ ....Generally distributed.
Columbus ♂ ....Upper Amazon, abundant.
Pausanias ....Upper Amazon, scarce.
Lycidas ......Upper Amazon, abundant.
Belus ....Upper Amazon, abundant.

b. Species frequenting gardens and open grounds :—

Thoas ....Abundant.
Torquatus .....Abundant.
Polydamas ....Common everywhere.
Caudius ......Scarce.

c. Species which rarely leave the forest :—

Species peculiar to the Upper Amazon.

Patros.	Ergeteles.	Cyamon.
Cixius.	Brissonius.	Evagoras.
Orellana.	Bolivar.	Chabrias.

Species occurring at Pará.

Diceros.	Hierocles.	Sesostris.
Æneides.	Marcius.	Hippason.
Æneas.	Aglaope.	Ariarathes.
Thelios.	Sonoria.	Anchisiades.
Parsodes.	Vertumnus.	Triopas.
Echelus.	Cutora.	

The *Pieridæ* generally are open-ground butterflies, two genera only, *Leptalis* and *Terias*, being true denizens of the forest. Of the former but one species, the *L. Eumelia*, is at all common. It, however, may be found almost everywhere, hovering slowly along in shady woods and plantations, and offering an easy prey to the Entomologist as well as to the more merciless insect-eating birds. The species of *Terias* prefer rather the dry and more open parts of the forest country, and often even come out into the full sunshine. Several species of *Pieris* accompany them; but they cannot vie in strength and rapidity of flight with the species of *Callidryas*, which rejoice in the hottest sunshine, and crowd in dense masses of several yards in extent around puddles and on sandy beaches, rising in clouds of yellow and orange on being disturbed.

The *Rhodoceras* again are of a sylvan taste, the handsome *R. Leachiana* being found only in open paths through the forest.

We now come to a group of insects peculiar to America, the *Ageronidæ*, and in them we first see a deviation from the normal manner of carrying the wings in repose; the species of this family invariably resting with the wings expanded. Five species are found about Pará, and they all frequent dry situations, and always settle on trunks of trees, with the head downwards. The singular noise produced by these insects has been noticed by Lacordaire and Mr. Darwin. The common species, *A. Feronia*, produces it remarkably loud, when two insects are chasing each other and constantly striking together. One alone does not produce the sound in flying; and I have never heard it made by the small species, *A. Chloe*, which is equally common with the other. I am inclined, therefore, to believe that it is produced in some way by the contact of two insects, and that only the larger and stronger winged species can produce it.

Like M. Lacordaire, I have found the pupa of a species of this family, but never the larva. It agreed exactly with his description, and was attached to a post, and braced like that of a *Papilio*.

Of the *Danaidæ* two species only are found on the Amazon. They frequent the most open situations, fly low, and constantly settle on herbaceous plants. In the beautiful family of *Heliconidæ*, the glory of South American Entomology, the Amazon valley is particularly rich, at least sixty or seventy species being found there, of which a considerable number seem peculiar. And here the same thing takes place which we observed with regard to the *Papilios*,--that the more rare and restricted species are those which inhabit the forest, while the species found in the open grounds are generally widely distributed, and often seem mere stragglers from other parts of the continent. Among these latter are *Lycorea Halia*, *Tithorea Megara*, and *Mechanitis Lysidice*, while most of the species of *Heliconia*, *Thyridia* and *Ithomia*, which prefer the forest shades, are confined to a comparatively limited district. The most characteristic of the Amazon valley are those species of *Heliconia* with white or yellow spots on a shining blue or black ground, such as the *Antiocha, Thamar*, and several others; those with radiating red lines on the lower wings, such as *Erythræa, Egeria, Doris*, and several undescribed species; and, lastly, the delicate little clear wings of the genera *Thyridia, Ithomia* and *Sais*. All these groups are exceedingly productive in closely allied species and varieties of the most interesting description, and often having a very limited range; and as there is every reason to believe that the banks of the lower Amazon are among the most recently formed parts of South America, we may fairly regard those insects, which are peculiar to that district, as among the youngest of species, the latest in the long series of modifications which the forms of animal life have undergone.

The *Heliconidæ* are the most elegant of butterflies, whether upon the wing or in a state of repose. Their bodies, their wings and their antennæ are all long, slender and well formed, and they are universally adorned with the most brilliant and harmonious colours. They fly rather slowly, and the little *Ithomias* hover almost invisible among the dark foliage. They all rest with their wings erect, upon leaves and flowers, and at night I have observed them asleep, hanging at the extreme end of a slender twig, which bends with their weight and swings gently with the evening breeze.

At the extremity of the abdomen these insects possess a

peculiar gland or appendage, concealed between the valves of the anus, but capable of being protruded. It is most developed in *Lycorea halia*, where it takes the form of a radiate tuft of hairs, forming, when exserted, two feathery globes at the extremity of the abdomen. In an undescribed species allied to *Heliconia Erythræa* it is also much developed, in the shape of small fleshy balls of an orange colour, which are always exserted when the insect is captured, and give out a penetrating aromatic odour something resembling chamomile.

As a whole, the *Heliconidæ* are the most abundant butterflies of the Amazon, and form by far the most striking feature in the Entomology of the country.

The species of *Acræa*, *Eneides* and *Eresia* remarkably resemble *Heliconidæ* in their mode of flight and habits, as well as in their form and markings, and they are almost always found in company with the species of that family which frequent open grounds, for none of them are forest insects.

The next genera, *Colænis*, *Agraulis*, *Melitæa*, *Euptoieta*, *Junonia* and *Anartia*, closely resemble our own *Nymphalidæ*. They frequent open grounds, delight in hot sunshine, and the Amazon species are all common and widely distributed. The larva of the *Agraulis Vanilla* feeds on the Passion-flower, as, Mr. Doubleday says, it does in Florida. *Colænis Dido* is a beautiful object in the sunshine, its wings appearing of a bright transparent green.

The species of *Eubagis* are delicate little butterflies, which frequent flowers and low herbage on the skirts of the forest. *Timetes, Marpesia* and *Victorina* are three genera, the species of which are often found in company, on the skirts of woods and at the water's edge, together with the *Callidryas. Timetes Chiron* and *Orsilochus* are common at Pará, in the more open forest; *Themistocles, Chrethon, Berania* and *Tutelina*, on the Upper Amazon, generally at the water's edge.

The beautiful genus *Cybdelis* is one almost peculiar to the great valley of the Upper Amazon. Scarcely a species is known from Pará, while nearly a dozen occur in the interior. They all fly with excessive rapidity, and are exceedingly shy; they settle on trunks of trees or on rocks by the water, where several species are only found. They swarm on the granite rocks of the Rio Negro in the heat of the day, where the places they sit upon would be hot enough for culinary operations.

The species of *Epicalia*, ornamented with the richest blue and orange bands upon a velvety black ground, are among the most brilliant of the diurnal Lepidoptera, and many of them grace the dense forests of the Amazon. Those beautiful insects, *Epicalia Ancea, E. Acontius* and *E. Numilius*, are found at Pará, while in the interior several others occur. They much resemble each other in their habits; they fly strongly, but with an even sailing motion, frequently moving in a circle, and returning repeatedly to the same leaf. They frequent shady glades in the damp parts of the forest, and usually sit with the wings erect, but will occasionally select a spot where a gleam of sunshine enters, and there expand their wings to enjoy its warmth. Many of these insects have been ascertained to be the males of species of the genus *Myscelia*, which differ remarkably from them in colour. Others, however, like *E. Ancea*, have females more nearly resembling themselves. The two groups should be kept separate.

The *Callitheas* are another genus of butterflies unsurpassed for exquisite beauty. The *C. Sapphira* inhabits the south bank of the Lower Amazon, while immediately opposite to it is found an allied species, the *C. Leprieurii*. A third species has been discovered by Mr. Bates, on the south side of the Upper Amazon, and has been named after him; while a fourth, equally beautiful, which I found near the sources of the Rio Negro, was lost with the rest of my collections on my return voyage, and so remains to be rediscovered. They all frequent dry, open woods. The females often fly low, but the males settle on the leaves of lofty trees, and are very difficult to capture. The *C. Leprieurii* was found plentifully on the trunks of trees, where a black sap was exuding. In their flight they much resemble the *Epicalias*.

The next group of insects I shall mention are but poorly represented in the Amazon valley. The *Catagrammas* seem to abound most in the mountains, and not more than three species occur in the district under consideration. As might be expected, therefore, they prefer dry situations, and are very active in their movements; they often settle on the ground and on trunks of trees, and after rain come to the open places about houses. The *Heterochroas*, the representatives of the *Limenitis* of the old world, resemble them in their steady sailing flight. They are found in the open glades and on the skirts of the forest.

Callizona Acesta and *Gynecia Dirce* are two singular butterflies, resembling one another in their markings and in their habits. They settle upon trunks of trees as invariably as the *Ageronias*, but always carry their wings in an erect position. The two are often seen together, and are both equally common. They frequent trees from which sap is exuding, and never by any chance rest upon a leaf or flower.

Prepona Demophron and *Hypna Clytemnestra* have very similar habits to the last, but are not so exclusively attached to a particular station; sitting upon a stick or leaf when it appears most convenient. They fly very strongly and delight in dry sunny places. The genera *Paphia* and *Siderone* have very similar habits, and frequent the same situations.

Aganisthos Orion, Megistanis Cadmus and *M. Bœotus* are very strong-winged insects, and are generally found near the rivers, preferring to settle on damp spots on the ground, and often at the water's edge, where in the dry season, on the Upper Amazon and Rio Negro, they are very abundant. The

M. Bœotus first occurs on the Amazon and Rio Negro, a few hundred miles above their junction, and continues abundant up to the Andes. The blue and the orange varieties are found together, and in almost equal numbers, and there is little doubt are sexes of one species.

We now come to the giants among butterflies, the *Morphidæ* and *Brassolidæ*. The magnificent *Morphos* are abundant at Pará and in every part of the Amazon valley; but the rare *Hecuba* is found only in the interior. The *Morphos* are forest insects, rarely coming into the open grounds, and often flying for miles along roads and open pathways. Their flight is slow and undulating, but they are very difficult to take on the wing. At Barra, on the Rio Negro, in the month of July, I found *Morpho Hecuba* very abundant. Its habits differed from the other species, as it frequented the open woods and glades in the vicinity of the city, but only for about an hour every morning from 9 to 10 o'clock; at other times not one was to be seen. It flew rather high, and was so very cautious in rising suddenly to escape the net, that, notwithstanding all my endeavours, I succeeded in obtaining very few specimens. These are all diurnal insects, while the genera, *Caligo* or *Pavonia*, *Dynastor*, *Opsiphanes* and *Brassolis* are truly crepuscular, never flying by day except when disturbed, but appearing to be voluntarily active only for about half an hour before sunrise and after sunset. They remain hid during the day in the gloomiest shades of the forest, where I have sometimes encountered them; and even when they come out they take but short flights, frequently settling either on the ground in a pathway, or on a leaf, before resting on which they generally make several trials to find one which will not bend too much with their weight. I have bred *Brassolis Sophoræ* from the larva which feeds on the leaves of the cocoa-nut palm and much disfigures them.

The next family, the *Satyridæ*, agrees with the last as much in habits as in form and marking. The great characteristic of the South American *Satyridæ* is their frequenting the shadiest parts of the forest, and their invariably flying along or very near the ground. So universal is this habit that I do not remember to have ever seen any species rise four feet from the earth, while the greater number of them do not exceed as many inches. They rest also upon the lowest herbage or upon the ground itself. The exquisite clear-winged *Hæteras* all have this habit in perfection. *H. Piera* is common all over the Amazon district. *Esmeralda* is widely distributed, but scarce; and *Andromeda* does not occur at Pará, but is found in company with *Esmeralda* in the interior. *H. Philoctetes* is also widely distributed, but very rare, and I took one specimen of the singular *Caerois Chorinæus* at Santarem, on the south or Brazilian bank of the Amazon, in a dry open forest.

I will here mention the beautiful *Bia Actorion*, which, though classed with the *Nymphalidæ*, exactly agrees with this family in its haunts and mode of flight; and as it

21

agrees with them also in many structural points, it may be considered as forming a very satisfactory link connecting the two families.

Didonis Biblis, belonging to the *Eurytelidæ*, is common in many parts of the Amazon. It frequents open and rather sunny places, but flies low and weakly.

A singular butterfly with greatly developed palpi, named *Libythea Cuvieri* in the British Museum Collection, I found abundantly at Santarem flying about marshy meadows in the sunshine.

We now come to the *Erycinidæ*, another extensive family of almost peculiarly American insects, and which exhibit a variety and brilliancy of colouring unsurpassed in the whole order. In these too the Amazon is particularly rich, producing about two hundred species. More than any other butterflies, the *Erycinidæ* are the inhabitants of the virgin forest in whose dark recesses many of the rarer and lovelier species are alone to be found. Some of the larger kinds, such as the genera *Helicopis*, *Erycina* and *Stalachtis* are common in the skirts of the forest, and even come out into the sunshine, where *Eurybia*, *Eurygona* and *Mesosemia* are scarcely ever seen.

The great mass of the species of this family have a very peculiar habit of invariably settling and reposing on the under surface of leaves with the wings expanded, but there are some striking exceptions to this rule. The beautiful golden-spotted *Helicopis cupido* and allied species, though they always settle on the under-sides of leaves, yet invariably do so with the wings closed, as in the more typical butterflies. The species of *Charis* and *Themone*, on the other hand, prefer the upper surface of leaves for their station, where they expand their wings; while the true *Erycinas* rest with wings closed upon leaves, like the typical *Papilios*, which by their tailed wings they distantly resemble, though their strong, rapid and jerking flight is like that of the *Hesperias*. These different modes of resting are very peculiar and interesting, because they are so very constant and uniform in the same genus. Is there any anatomical peculiarity which leads *Nymphidium* always to expand its wings in repose and *Helicopis* always to close them? and for what reason should *Charis* always expose itself on the upper surfaces of leaves, while hundreds of its more modest or more timid allies invariably take advantage of the friendly shelter afforded them and rest upon the under side?

There are some other peculiarities in the habits of the different genera worthy of notice. *Eurybia* and *Eunogyra* always keep within a few inches of the ground, like the *Satyridæ*, which in their peculiar ocellated markings they both so much resemble; almost leading us to suppose that the colour and marking of an insect affects its habits, or vice versâ. The *Erycinas*, as I have already remarked, fly as strongly and as rapidly as the *Hesperias*. *Emesis* and *Nymphidium* are also rapid flyers, while *Stalachtis*, which approaches *Heliconia* in size and markings, resembles it also in its flight and frequents the same situations. Most of the other genera are weak but rather active flyers, and from their small size and their suddenly disappearing beneath a leaf, are not always easy to capture.

We now come to the *Theclas*, of which about sixty species may be found at Pará, more than half of which are still undescribed. *Endymion*, *Marsyas*, *Etolus* and others, frequent the open grounds in the sunshine, while *Pholeus* and many more are found only in the depths of the forest. All fly very quickly and settle upon leaves and flowers with the wings erect. They have a very peculiar habit of moving the two lower wings over each other in opposite directions, giving an appearance of revolving discs. In the interior the species are not nearly so numerous as at Pará, and this seems to be generally the case in other parts of Brazil, many species being recorded from Pernambuco and Rio as well as from Honduras and Florida, but few from the mountainous districts in the interior of the country. I do not find a single species recorded from Peru or Bolivia or any of the countries west of the Andes.

We have now reached the last family of butterflies, the *Hesperidæ*, which are particularly abundant in South America, and of which about 200 species may be taken at Pará. Like the *Theclas*, they are far less numerous in the interior. Our own little skippers afford but a faint idea of the variety and beauty of the tropical *Hesperidæ*, some of which expand upwards of three inches. Neither do the habits of our species hold good for the whole family, which presents many important characters. There are three distinct modes in which the wings are carried in repose:--1st. They are closed and carried erect as in the typical butterflies; 2nd. The fore wings are elevated while the hind ones expand; and 3rd. The wings are all expanded. The great majority of the South American species belong to the first of these divisions, such as the genera *Pyrrhopyga, Ericides, Goniurus, Goniloba*, and some species of *Pamphila*, as *P. Epictetus*. To the second division, which includes one British species, very few South American *Hesperidæ* belong, principally of the genus *Pamphila*; but the third comprehends a very peculiar group of insects, consisting of the genera *Pyrgus, Nisoniades* and *Achilodes*. They have the upper wings more or less convex, and never erect them in repose, and they will, I believe, form a very natural subdivision of the family.

The different genera vary much in the situations they frequent. The beautiful *Pyrrhopyga Mænas* and *Ericides Gnetus* were found in open places on sandy beaches, often settling on the ground and seeking out dead fish or decaying animal matter to alight upon. *Goniuris Proteus* prefers open grounds in the hot sunshine, as does *Goniloba Exadeus*, while the greater number of the species with vitreous spots occur in the thick woods and many in the depths of the virgin forest.

The smaller species have the characteristic mode of flight which has obtained for them the name of skippers, but some of the larger and strong-bodied kinds are remarkable for the excessive rapidity of their flight, which I believe exceeds that of any other insects. The eye cannot follow them as they dart past; and the air, forcibly divided, gives out a deep sound louder than that produced by the humming-bird itself. If power of wing and rapidity of flight could place them in that rank, they should be considered the most highly organized of butterflies.

Several species of *Castnia* occur on the Amazon, which, though diurnal in their habits and brightly coloured, cannot be classed with the butterflies. They also rest with their wings deflexed, so that the upper only are visible, after the manner of the *Bombycidæ*, and they generally sit upon the end of a stick or twig in the full sunshine.

In conclusion, I venture to hope that if my observations are wanting in detail and in precision as to the exact species to which they apply, it may be imputed, not to want of accuracy on my part, but to the loss of a large portion of my notes and collections during my return voyage to this country. I trust, however, that in the absence of much information on the habits of exotic insects, my remarks, however imperfect, may not be altogether valueless.

www.ingramcontent.com/pod-product-compliance
Lightning Source LLC
Chambersburg PA
CBHW021340290326
41933CB00038B/995